Fractions: Concepts and Probl...
Grade 5

Table of Contents

Fractions: Concepts and Problem Solving
Grade 5

Introduction

Helping students form an understanding of fractions is a challenging process. To best accomplish the task, educators must approach the subject in a way that is meaningful for students. Moreover, the National Council of Teachers of Mathematics (NCTM) has indicated that "students should build their understanding of fractions as parts of a whole and as division." (*Principles and Standards for School Mathematics,* page 150.) *Fractions: Concepts and Problem Solving* serves as a companion to the classroom mathematics curriculum and encompasses many of the standards established by the NCTM for this grade level. (Refer to the correlation chart below.) The book is divided into four units: Concept Development, Computation, Problem Solving, and Enrichment. Each page targets a specific skill to help bolster students who need additional work in a particular area.

Unit 1: Concept Development
The pages in this unit use a variety of learning styles to help students understand the fundamental principles of fractions. Students identify fractions and mixed numbers. They also develop skills to compute fractions, such as finding the least common multiple, the greatest common factor, equivalent fractions, and simplest form. They also compare and order fractions.

Unit 2: Computation
Here, students are introduced to the steps necessary to calculate fractional algorithms using visuals. Practice exercises on most pages guide students through each step. Students add, subtract, multiply, and divide fractions with like and unlike denominators.

Unit 3: Problem Solving
Students work word and real-life application problems to further develop skills in fractions. They work with patterns, bar graphs, line graphs, and circle graphs to find fractional relationships. They also explore basic probability.

Unit 4: Enrichment
To challenge and extend learning, students explore basic algebra principles, decimals, ratios, percents, and money.

Special Note
We encourage the use of manipulatives for acquisition of skills with fractions. Examples include, but are not limited to, the following: Fraction Bars®, Fraction Tiles®, Fraction Builder® Strips, Rainbow Fraction Circles and Squares®, Fraction Stax®, and Fraction Burger®.

Notes

Assessment
There are two kinds of assessment pages.
- On pages 5 and 6 is a general assessment that covers important fraction skills appropriate for fifth grade. It can be given as a pretest to gauge students' knowledge of fractions. Later in the year, the same test can be administered to determine students' understanding, progress, and achievement.
- The first three units also have an assessment. They can be administered at any time during the unit as a pretest, review, or posttest for specific fraction concepts.

NCTM Standards Correlation
The NCTM Standards Correlation chart below identifies a variety of mathematics standards that are appropriate for the study of fractions.

Fraction Table
A Fraction Table can be found on page 3. Students can use the graphic organizer to quickly identify and compare fractions. You may wish to make 2 photocopies for each student. One copy can be cut apart and used as manipulatives.

Fraction Circles
Fraction circles representing basic fractions are on page 4. Make a copy for each student. Students can cut them apart and use the circles as manipulatives while they work. You may wish to provide envelopes for students to store cut pieces.

NCTM Standards Correlation

Numbers and Operations: 7, 8, 9, 10, 11, 12, 13, 14, 15, 16, 17, 18, 19, 20, 21, 22, 23, 24, 25, 26, 27, 28, 29, 30, 31, 32, 33, 34, 35, 36, 37, 38, 39, 40, 41, 42, 44, 45, 46

Algebra: 7, 9, 10, 11, 14, 30, 31, 40, 41

Geometry: 7, 8, 9, 13, 16, 18, 30, 39

Measurement: 9, 10, 11, 28, 29, 31, 32, 33, 34, 37, 38, 46

Data Analysis and Probability: 28, 32, 34, 35, 36, 37, 38, 44, 45

Problem Solving: 9, 19, 28, 29, 30, 31, 32, 33, 35, 36, 37, 38, 39

Name ______________________ Date ______________

Fraction Table

1											
$\frac{1}{2}$	$\frac{1}{2}$										
$\frac{1}{3}$	$\frac{1}{3}$	$\frac{1}{3}$									
$\frac{1}{4}$	$\frac{1}{4}$	$\frac{1}{4}$	$\frac{1}{4}$								
$\frac{1}{5}$	$\frac{1}{5}$	$\frac{1}{5}$	$\frac{1}{5}$	$\frac{1}{5}$							
$\frac{1}{6}$	$\frac{1}{6}$	$\frac{1}{6}$	$\frac{1}{6}$	$\frac{1}{6}$	$\frac{1}{6}$						
$\frac{1}{7}$	$\frac{1}{7}$	$\frac{1}{7}$	$\frac{1}{7}$	$\frac{1}{7}$	$\frac{1}{7}$	$\frac{1}{7}$					
$\frac{1}{8}$	$\frac{1}{8}$	$\frac{1}{8}$	$\frac{1}{8}$	$\frac{1}{8}$	$\frac{1}{8}$	$\frac{1}{8}$	$\frac{1}{8}$				
$\frac{1}{9}$	$\frac{1}{9}$	$\frac{1}{9}$	$\frac{1}{9}$	$\frac{1}{9}$	$\frac{1}{9}$	$\frac{1}{9}$	$\frac{1}{9}$	$\frac{1}{9}$			
$\frac{1}{10}$	$\frac{1}{10}$	$\frac{1}{10}$	$\frac{1}{10}$	$\frac{1}{10}$	$\frac{1}{10}$	$\frac{1}{10}$	$\frac{1}{10}$	$\frac{1}{10}$	$\frac{1}{10}$		
$\frac{1}{11}$	$\frac{1}{11}$	$\frac{1}{11}$	$\frac{1}{11}$	$\frac{1}{11}$	$\frac{1}{11}$	$\frac{1}{11}$	$\frac{1}{11}$	$\frac{1}{11}$	$\frac{1}{11}$	$\frac{1}{11}$	
$\frac{1}{12}$	$\frac{1}{12}$	$\frac{1}{12}$	$\frac{1}{12}$	$\frac{1}{12}$	$\frac{1}{12}$	$\frac{1}{12}$	$\frac{1}{12}$	$\frac{1}{12}$	$\frac{1}{12}$	$\frac{1}{12}$	$\frac{1}{12}$

Fraction Circles

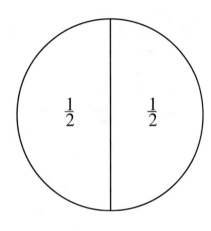

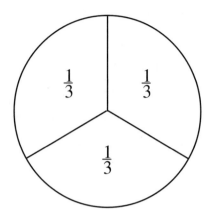

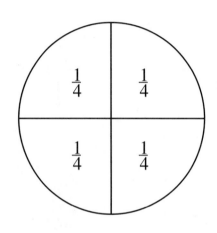

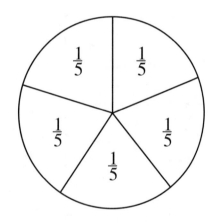

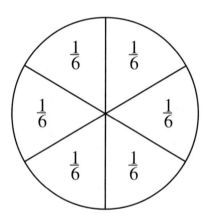

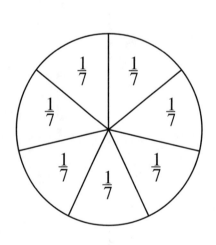

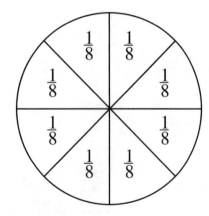

General Assessment

Directions

Darken the circle by the correct answer.

1. $\frac{5}{6}$ is equal to _____.

   Ⓐ $\frac{5}{10}$

   Ⓑ $\frac{10}{12}$

   Ⓒ $\frac{5}{12}$

   Ⓓ $\frac{1}{10}$

2. Which is true?

   Ⓐ $\frac{1}{4} > \frac{1}{3}$

   Ⓑ $\frac{4}{7} < \frac{2}{5}$

   Ⓒ $\frac{2}{9} > \frac{1}{3}$

   Ⓓ $\frac{7}{10} > \frac{4}{15}$

3. Reduce $\frac{5}{60}$ to its simplest form.

   Ⓐ $\frac{1}{15}$

   Ⓑ $\frac{5}{12}$

   Ⓒ $\frac{6}{15}$

   Ⓓ $\frac{1}{12}$

4. What fraction shows how many of this group are <u>not</u> shaded?

   Ⓐ $\frac{3}{10}$

   Ⓑ $\frac{7}{3}$

   Ⓒ $\frac{7}{10}$

   Ⓓ $\frac{3}{7}$

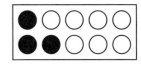

5. What is the least common multiple of 3 and 4?

   Ⓐ 6

   Ⓑ 8

   Ⓒ 10

   Ⓓ 12

6. What is the greatest common factor of 8 and 12?

   Ⓐ 2

   Ⓑ 4

   Ⓒ 6

   Ⓓ 8

7. Which number renames $\frac{17}{8}$?

   Ⓐ $2\frac{1}{8}$

   Ⓑ 3

   Ⓒ $2\frac{5}{8}$

   Ⓓ 2

8. Which mixed number does the picture stand for?

   Ⓐ $3\frac{6}{8}$

   Ⓑ $3\frac{1}{8}$

   Ⓒ $4\frac{1}{8}$

   Ⓓ $4\frac{7}{8}$

Go on to the next page.

General Assessment, p. 2

9. $\frac{1}{10}$
  $+ \frac{3}{5}$

   Ⓐ $\frac{7}{10}$

   Ⓑ $\frac{6}{10}$

   Ⓒ $\frac{6}{20}$

   Ⓓ $\frac{8}{20}$

10. $\frac{3}{4} \times 8 =$ _____

   Ⓐ $4\frac{1}{2}$

   Ⓑ 6

   Ⓒ $2\frac{4}{8}$

   Ⓓ 11

11. $9\frac{1}{3} \times \frac{2}{6} =$ _____

   Ⓐ $2\frac{2}{9}$

   Ⓑ $3\frac{2}{9}$

   Ⓒ $3\frac{1}{8}$

   Ⓓ $3\frac{1}{9}$

12. $\frac{1}{3} \div \frac{3}{9} =$ _____

   Ⓐ $\frac{5}{12}$

   Ⓑ $\frac{9}{12}$

   Ⓒ $\frac{3}{4}$

   Ⓓ 1

13. Which is the fraction for 20%?

   Ⓐ $\frac{3}{5}$

   Ⓑ $\frac{4}{5}$

   Ⓒ $\frac{1}{5}$

   Ⓓ $\frac{2}{5}$

14. A bag contains 5 clear marbles and 10 colored marbles. What is the probability of reaching into the bag and pulling out a clear marble?

   Ⓐ $\frac{10}{15}$

   Ⓑ $\frac{7}{15}$

   Ⓒ $\frac{5}{15}$

   Ⓓ $\frac{3}{15}$

15. Sam and Jan went fishing. Sam caught a trout that weighed $3\frac{1}{2}$ pounds. Jan caught a trout that weighed $4\frac{1}{4}$ pounds. How much heavier is Jan's trout than Sam's?

   Ⓐ $\frac{3}{4}$ pound

   Ⓑ $1\frac{1}{2}$ pounds

   Ⓒ $\frac{1}{4}$ pound

   Ⓓ $\frac{1}{2}$ pound

16. George ate $\frac{1}{3}$ of a pizza. His friend Jack ate $\frac{1}{6}$ of the pizza. How much of the pizza was eaten?

   Ⓐ $\frac{2}{6}$ pizza

   Ⓑ $\frac{2}{3}$ pizza

   Ⓒ $\frac{1}{2}$ pizza

   Ⓓ $\frac{2}{9}$ pizza

Name _____ Date _____

Unit 1 Assessment

Directions

Write the fraction for the part that is shaded.

1.

_____

2.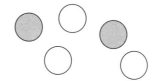

_____

Directions

Compare. Write $<$, $>$, or $=$.

3. $\frac{1}{6}$ ◯ $\frac{5}{6}$

4. $\frac{3}{4}$ ◯ $\frac{1}{8}$

5. $\frac{3}{10}$ ◯ $\frac{4}{5}$

Directions

Write each fraction in its simplest form.

6. $\frac{4}{10}$ _____

7. $\frac{3}{8}$ _____

8. $\frac{6}{12}$ _____

Directions

Write the greatest common factor for each pair of numbers.

9. 6 and 8
GCF _____

10. 9 and 12
GCF _____

11. 4 and 16
GCF _____

Directions

Arrange in order from least to greastest.

12. $\frac{1}{4}, \frac{7}{8}, \frac{1}{3}$

_____

13. $2\frac{1}{9}, 2\frac{3}{5}, 4\frac{1}{9}$

_____

Understanding Fractions

A **fraction** can name part of a whole or part of a group.

Part of a whole.	**Part of a group.**

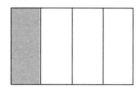 

The rectangle is divided into 4 equal parts. There is 1 shaded part.

There are 5 circles. There are 2 circles shaded.

number of shaded parts →1← numerator
number of parts in all →4← denominator

number of circles shaded →2← numerator
number of circles in all →5← denominator

So, $\frac{1}{4}$ of the rectangle is shaded. Read the fraction as one fourth, one out of four, or one divided by four.

So, $\frac{2}{5}$ of the circles are shaded. Read the fraction as two fifths, two out of five, or two divided by five.

Directions

Write the fraction for the shaded part.

1.

2.

3.

_____   _____   _____

Directions

Shade the correct number of parts to show the fraction.

4. $\frac{2}{3}$

5. $\frac{1}{3}$

6. $\frac{5}{6}$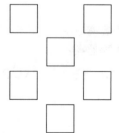

Name _____ Date _____

Estimating Fractions

You can use a fraction bar to help you **estimate** fractions.

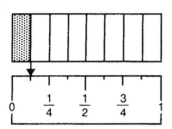

$\frac{1}{8}$ is about 0.

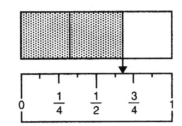

$\frac{2}{3}$ is about $\frac{1}{2}$.

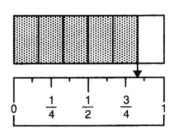

$\frac{5}{6}$ is about 1.

(Directions)

Use the fraction bar above to estimate the shaded part of each. Write *about 0, about $\frac{1}{2}$,* or *about 1.*

1.
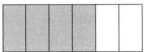

$\frac{4}{6}$ _____

2.
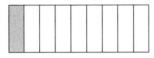

$\frac{1}{9}$ _____

3.

$\frac{8}{9}$ _____

(Directions)

Circle the better estimate.

4.
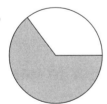

$\frac{5}{8}$ or $\frac{7}{8}$

5.

$\frac{1}{2}$ or $\frac{2}{3}$

6.
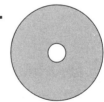

$\frac{3}{4}$ or $\frac{1}{2}$

Name _____   Date _____

Equivalent Fractions

You can use different fractions to name the same amount. Fractions that name the same amount are called **equivalent fractions**. There are several ways to find equivalent fractions.

The shaded part is $\frac{1}{2}$ of the rectangle.
The shaded part is also $\frac{2}{4}$ of the rectangle.
So, $\frac{1}{2}$ and $\frac{2}{4}$ are two names for the same fraction.

$\frac{1}{2} = \frac{2}{4}$

You can find equivalent fractions in three ways.

Use a number line.	Multiply both the numerator and the denominator by the same number.	Divide both the numerator and the denominator by the same number.
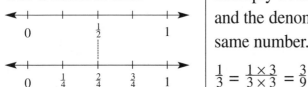	$\frac{1}{3} = \frac{1 \times 3}{3 \times 3} = \frac{3}{9}$	$\frac{6}{8} = \frac{6 \div 2}{8 \div 2} = \frac{3}{4}$
You can see that $\frac{1}{2} = \frac{2}{4}$, so they are equivalent fractions.	The fraction $\frac{1}{3}$ names the same amount as $\frac{3}{9}$, so they are equivalent fractions.	The fractions $\frac{6}{8}$ and $\frac{3}{4}$ are equal, so they are equivalent fractions.

Directions Use the number line to find out if the fractions are equivalent. Write *yes* or *no*.

1. $\frac{1}{4} = \frac{3}{12}$ _____

2. $\frac{8}{12} = \frac{3}{4}$ _____

Directions Multiply both the numerator and the denominator to name an equivalent fraction.

3. $\frac{3}{8} = \frac{3 \times 2}{8 \times 2} = \dfrac{\square}{\square}$

4. $\frac{2}{3} = \frac{2 \times 5}{3 \times 5} = \dfrac{\square}{\square}$

Directions Divide both the numerator and the denominator to name an equivalent fraction.

5. $\frac{12}{16} = \frac{12 \div 4}{16 \div 4} = \dfrac{\square}{\square}$

6. $\frac{7}{28} = \frac{7 \div 7}{28 \div 7} = \dfrac{\square}{\square}$

Name _____ Date _____

Least Common Multiple

The product of two or more numbers is a **multiple**. Multiples of one number that are also multiples of another number are called **common multiples**. The smallest number that is a common multiple is called the **least common multiple**, or **LCM**.

Example: Find the least common multiple of 4 and 6.

Step 1: Find the multiples of the numbers. A number line can help.

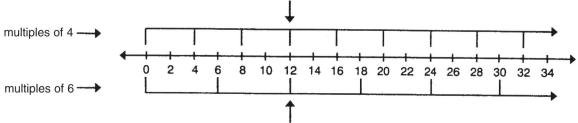

Step 2: Circle the first multiple the numbers have in common.

Multiples of 4: 4, 8, (12,) 16, 20, 24, 28, 32, 36, ...
Multiples of 6: 6, (12,) 18, 24, 30, 36, ...

So, the least common multiple of 4 and 6 is 12.

Directions

Use the number line to find the least common multiple for each pair of numbers. Write the least common multiple.

1. 4 and 5 _____   **2.** 6 and 8 _____   **3.** 5 and 10 _____

Directions

List 5 multiples of each number. Then, find the least common multiple.

4. 3 _____   **5.** 3 _____   **6.** 3 _____
   4 _____   5 _____   6 _____
 LCM _____   LCM _____   LCM _____

Name _____ Date _____

Greatest Common Factor

The greatest factor that two or more numbers have in common is the **greatest common factor**, or **GCF**.

Example: Find the greatest common factor of 8 and 12.

Step 1
Find the prime number factors of each number. A factor tree can help.

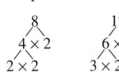

Step 2
Find the factors that are common to both numbers.
$8 = 2 \times \boxed{2} \times \boxed{2}$
$12 = 3 \times \boxed{2} \times \boxed{2}$

Step 3
Multiply the common factors.
$2 \times 2 = 4$

So, the greatest common factor of 8 and 12 is 4.

Directions

Follow the steps to find the greatest common factor for 6 and 18.

1. Step 1

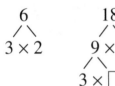

2. Step 2
$6 = 3 \times 2$
$18 = 3 \times \boxed{} \times \boxed{}$

3. Step 3
$\boxed{} \times \boxed{} = \boxed{}$
The GCF is _____.

Directions

Find the GCF. Use another sheet of paper.

4. 4 and 8
GCF _____

5. 9 and 12
GCF _____

6. 8 and 16
GCF _____

Unit 1: Concept Development
Fractions 5, SV 3408-8

Name _____   Date _____

Simplest Form

A fraction is in **simplest form** when the greatest common factor (GCF) of the numerator and denominator is 1.

Example: Write $\frac{4}{8}$ in its simplest form.

Step 1
Find the factors of each number.

Step 2
Determine the GCF.
$4 = \boxed{2} \times \boxed{2}$
$8 = \boxed{2} \times \boxed{2} \times 2$
$2 \times 2 = 4$
The GCF is 4.

Step 3
Divide the numerator and the denominator by the GCF.
$\frac{4}{8} = \frac{4 \div 4}{8 \div 4} = \frac{1}{2}$.

So, $\frac{4}{8}$ written in simplest form is $\frac{1}{2}$.

Directions

Follow the steps above to write each fraction in its simplest form. The first one is partly done for you.

1. $\frac{6}{8}$

6   $6 = 3 \times 2$ _____

8   $8 = 2 \times 2 \times 2$ _____

GCF _____

$\frac{6}{8} = \frac{6 \div \square}{8 \div \square} = \frac{\square}{\square}$

2. $\frac{14}{21}$

14 _____

21 _____

GCF _____

$\frac{14}{21} = \frac{14 \div \square}{21 \div \square} = \frac{\square}{\square}$

Directions

Write in simplest form.

3. $\frac{9}{15}$ _____   **4.** $\frac{10}{14}$ _____   **5.** $\frac{12}{18}$ _____

Comparing Fractions

When comparing fractions, look at the denominators. Fractions with unlike denominators are treated differently than fractions with like denominators.

Example 1: Compare $\frac{2}{5}$ and $\frac{3}{5}$. $\frac{2}{5}$  $\frac{3}{5}$

When comparing fractions with like denominators, the fraction with the greater numerator is the greater fraction.

$3 > 2$ So, $\frac{3}{5} > \frac{2}{5}$.

$\frac{2}{3}$ $\frac{1}{2}$

Example 2: Compare $\frac{2}{3}$ and $\frac{1}{2}$.

Step 1: Find the LCM of the denominators.   2: 2, 4,⑥,8, ...
3: 3,⑥,9, ...   6 is the LCM.

Step 2: Write equivalent fractions using the LCM. $\frac{2 \times 2}{3 \times 2} = \frac{4}{6}$ $\frac{1 \times 3}{2 \times 3} = \frac{3}{6}$

Step 3: Compare the numerators of the equivalent fractions you wrote. $4 > 3$ So, $\frac{2}{3} > \frac{1}{2}$.

Directions

Compare. Write the greater fraction in each pair.

1. $\frac{1}{4}, \frac{2}{4}$ _____   2. $\frac{2}{3}, \frac{1}{3}$ _____   3. $\frac{5}{8}, \frac{7}{8}$ _____

4. $\frac{1}{4}, \frac{1}{8}$ _____   5. $\frac{1}{2}, \frac{1}{3}$ _____   6. $\frac{2}{6}, \frac{3}{5}$ _____

Directions

Compare. Write <, >, or =.

7. $\frac{3}{4} \bigcirc \frac{2}{5}$ 8. $\frac{4}{5} \bigcirc \frac{5}{6}$

9. $\frac{2}{3} \bigcirc \frac{4}{6}$ 10. $\frac{1}{5} \bigcirc \frac{2}{10}$

11. $\frac{3}{7} \bigcirc \frac{4}{9}$ 12. $\frac{5}{12} \bigcirc \frac{3}{9}$

Name _____   Date _____

Ordering Fractions

> Sometimes you will need to order a set of numbers. There are several ways to find the order.

Example: Order the fractions $\frac{1}{2}$, $\frac{3}{4}$, $\frac{7}{8}$ from least to greatest. You can find the order of fractions in two ways.

Use fraction bars.

1							
$\frac{1}{2}$				$\frac{1}{2}$			
$\frac{1}{4}$		$\frac{1}{4}$		$\frac{1}{4}$		$\frac{1}{4}$	
$\frac{1}{8}$	$\frac{1}{8}$	$\frac{1}{8}$	$\frac{1}{8}$	$\frac{1}{8}$	$\frac{1}{8}$	$\frac{1}{8}$	$\frac{1}{8}$

So, $\frac{1}{2} < \frac{3}{4} < \frac{7}{8}$.

Use equivalent fractions.

Step 1: Find the LCM of 2, 4, and 8.

2: 2, 4, 6, ⑧, ...
4: 4, ⑧ 12, 16, ...   8 is the LCM.
8: ⑧ 16, 24, 32, ...

Step 2: Write equivalent fractions using the LCM.

$\frac{1 \times 4}{2 \times 4} = \frac{4}{8}$ $\frac{3 \times 2}{4 \times 2} = \frac{6}{8}$ $\frac{7}{8} = \frac{7}{8}$

Step 3: Compare the numerators.

$4 < 6 < 7$ So, $\frac{1}{2} < \frac{3}{4} < \frac{7}{8}$.

(Directions)

Write in order from least to greatest.

1. $\frac{1}{3}, \frac{1}{2}, \frac{1}{4}$ _____

2. $\frac{1}{3}, \frac{1}{4}, \frac{1}{6}$ _____

3. $\frac{1}{4}, \frac{1}{5}, \frac{1}{10}$ _____

4. $\frac{1}{2}, \frac{1}{4}, \frac{3}{8}$ _____

5. $\frac{1}{3}, \frac{5}{6}, \frac{4}{9}$ _____

6. $\frac{2}{3}, \frac{3}{4}, \frac{7}{12}$ _____

Mixed Numbers

A **mixed number** is made up of a whole number and a fraction. The fraction bars show the mixed number $3\frac{2}{3}$.

An **improper fraction** is a fraction where the numerator is greater than the denominator. The fraction is greater than 1. The fraction bars show the improper fraction $\frac{11}{3}$.

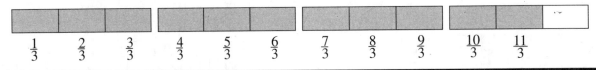

$\frac{1}{3}$ $\frac{2}{3}$ $\frac{3}{3}$ $\frac{4}{3}$ $\frac{5}{3}$ $\frac{6}{3}$ $\frac{7}{3}$ $\frac{8}{3}$ $\frac{9}{3}$ $\frac{10}{3}$ $\frac{11}{3}$

You can rename an improper fraction as a mixed number.

Example: Rename $\frac{11}{3}$.

Step 1:
Divide the numerator by the denominator.

$$\begin{array}{r} 3 \\ 3\overline{)11} \\ -9 \\ \hline 2 \end{array}$$

Step 2:
Show how many thirds are left over.

$$\begin{array}{r} 3\frac{2}{3} \leftarrow \text{remainder} \\ \overset{}{\leftarrow} \text{divisor} \\ 3\overline{)11} \\ -9 \\ \hline 2 \end{array}$$

So, $\frac{11}{3} = 3\frac{2}{3}$.

Directions

Write the mixed number.

1.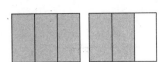

2.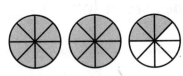

_____   _____

Directions

Rename each fraction as a mixed number.

3. $\frac{5}{4}$ 4. $\frac{9}{5}$ 5. $\frac{10}{6}$

_____   _____   _____

Name **Janie Williams**   Date **January 16, 2010**

Unit 2 Assessment

Directions

Add or subtract. Reduce to simplest form.

1. $\dfrac{5}{6}$
    $+\dfrac{2}{6}$
    $\dfrac{7}{6}$

2. $\dfrac{4}{7}$
    $-\dfrac{1}{7}$
    $\dfrac{3}{7}$

3. $\dfrac{1}{2}$
    $+\dfrac{2}{5}$
    $\dfrac{9}{10}$

4. $3\dfrac{4}{6}$
    $-2\dfrac{2}{6}$
    $1\dfrac{2}{6}$

5. $\dfrac{3}{7}$
    $+\dfrac{1}{2}$
    $\dfrac{13}{14}$

6. $4\dfrac{3}{5}$
    $+2\dfrac{1}{10}$
    $6\dfrac{7}{10}$

7. $5\dfrac{3}{4}$
    $-2\dfrac{3}{8}$

8. $2\dfrac{2}{3}$
    $+\dfrac{4}{6}$

Directions

Multiply or divide. Reduce to simplest form.

9. $\dfrac{1}{2} \times 2 = $ _____

10. $\dfrac{2}{3} \times \dfrac{1}{4} = $ _____

11. $\dfrac{5}{6} \div \dfrac{1}{2} = $ _____

12. $\dfrac{2}{3} \div 3 = $ _____

13. $\dfrac{3}{5} \times 1\dfrac{1}{2} = $ _____

14. $\dfrac{1}{4} \times 2\dfrac{2}{3} = $ _____

15. $\dfrac{3}{4} \div 6 = $ _____

Name Jania Williams   Date 1.16.10

Estimating Sums and Differences

When adding and subtracting fractions, it is sometimes helpful to **estimate** their values. The sign ≈ means to estimate the sum or difference.

Example: Estimate $\frac{5}{12} + \frac{8}{12} \approx \square$

    Step 1: Estimate the value of each fraction as closer to 0, $\frac{1}{2}$, or 1.

      $\frac{5}{12}$ is closer to $\frac{1}{2}$.
      $\frac{8}{12}$ is closer to $\frac{1}{2}$.

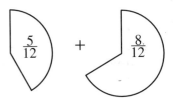

    Step 2: Rewrite the number sentence using the estimated values.

      $\frac{1}{2} + \frac{1}{2} = 1$

So, $\frac{5}{12} + \frac{8}{12} \approx 1$.

Directions

Estimate whether each fraction is closer to 0, $\frac{1}{2}$, or 1.

1. 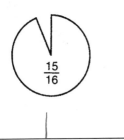 $\frac{15}{16}$

    |

2. $\frac{1}{8}$

    0

3. 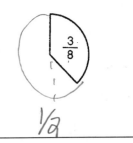 $\frac{3}{8}$

    1/2

4. $\frac{1}{20}$ 0

5. $\frac{7}{12}$ |

6. $\frac{9}{10}$ |

Directions

Estimate the sum or difference. You can use fraction pieces to help you.

7. $\frac{3}{8} + \frac{7}{16} \approx \frac{10}{24}$

8. $\frac{1}{8} + \frac{15}{16} \approx \frac{16}{24}$

9. $\frac{7}{8} - \frac{13}{16} \approx$ _____

10. $\frac{15}{16} - \frac{3}{8} \approx$ _____

11. $\frac{3}{8} + \frac{1}{16} \approx$ _____

12. $\frac{9}{16} - \frac{3}{8} \approx$ _____

Name  Janig Williams   Date _____

Adding and Subtracting Fractions with Like Denominators

Like fractions are fractions that have the same denominator. Since $\frac{3}{6}$ and $\frac{1}{6}$ both have denominator 6, they are **like fractions**. When the denominators are the same, it is easy to add or subtract the fractions.

Example 1: $\frac{2}{6} + \frac{2}{6} = \square$

    Step 1: Add the numerators. $2 + 2 = 4$

    Step 2: Write the same denominator. $\frac{4}{6}$

    Step 3: Reduce to simplest form. $\frac{4}{6} = \frac{4 \div 2}{6 \div 2} = \frac{2}{3}$

So, $\frac{2}{6} + \frac{2}{6} = \frac{2}{3}$.

Example 2: $\frac{6}{8} - \frac{2}{8} = \square$

    Step 1: Subtract the numerators. $6 - 2 = 4$

    Step 2: Write the same denominator. $\frac{4}{8}$

    Step 3: Reduce to simplest form. $\frac{4}{8} = \frac{4 \div 4}{8 \div 4} = \frac{1}{2}$

So, $\frac{6}{8} - \frac{2}{8} = \frac{1}{2}$.

Directions

Add or subtract. Reduce to simplest form.

1. $\frac{1}{6} + \frac{2}{6} = \dfrac{\boxed{3}}{6} = \dfrac{\square}{6} = \dfrac{\square}{\square}$

2. $\frac{7}{9} - \frac{4}{9} = \dfrac{\boxed{}}{9} = \dfrac{\square}{9} = \dfrac{\square}{\square}$

3. $\frac{1}{6} + \frac{2}{6}$

4. $\frac{3}{5} - \frac{2}{5}$

5. $\frac{1}{3} + \frac{2}{3}$

6. $\frac{9}{12} - \frac{1}{12}$

_____   _____   _____   _____

7. $\frac{4}{9} - \frac{2}{9}$

8. $\frac{5}{12} + \frac{11}{12}$

9. $\frac{6}{8} - \frac{4}{8}$

10. $\frac{5}{8} + \frac{7}{8}$

_____   _____   _____   _____

Adding Fractions with Unlike Denominators

When adding fractions, remember to look at the denominators first. If they are not the same, find a **common denominator**.

Example: Add $\frac{3}{4} + \frac{1}{3} = \boxed{}$

Step 1: Find a common denominator. Choose the least common multiple.

multiples of 4: 4, 8, ⑫, 16
multiples of 3: 3, 6, 9, ⑫, 15
LCM: 12

Step 2: Rename each fraction using the common denominator.

$\frac{3}{4} = \frac{9}{12}$ $\frac{1}{3} = \frac{4}{12}$

Step 3: Add the like fractions.

$\frac{9}{12} + \frac{4}{12} = \frac{13}{12}$

Step 4: Rename improper fractions and write the sum in simplest form.

$\frac{13}{12} = 1\frac{1}{12}$

Directions

Find a common denominator.

1. $\frac{1}{2} + \frac{2}{3}$

2. $\frac{3}{4} + \frac{1}{2}$

3. $\frac{3}{5} + \frac{1}{2}$

4. $\frac{1}{4} + \frac{5}{6}$

_____ _____ _____ _____

Directions

Add. Reduce to simplest form.

5. $\frac{1}{5} + \frac{1}{2}$

6. $\frac{2}{3} + \frac{3}{4}$

7. $\frac{2}{6} + \frac{1}{4}$

8. $\frac{1}{2} + \frac{5}{6}$

_____ _____ _____ _____

9. $\frac{3}{5} + \frac{3}{10}$

10. $\frac{1}{4} + \frac{5}{6}$

11. $\frac{3}{8} + \frac{3}{4}$

12. $\frac{1}{2} + \frac{1}{3}$

_____ _____ _____ _____

Name _____ Date _____

Adding Mixed Numbers

When adding mixed numbers, first add the fractions and then add the whole numbers. Sometimes, the sum in the numerator is greater than the denominator. This means the fraction is greater than 1. Be sure to rename the improper fraction as a mixed number.

Example: Add $3\frac{6}{8} + 2\frac{1}{2} = \square$

Step 1:
Rename each fraction using a common denominator.

$3\frac{6}{8} \rightarrow 3\frac{6}{8}$
$+ 2\frac{1}{2} \rightarrow 2\frac{4}{8}$

Step 2:
Add the fractions.
Add the whole numbers.

$3\frac{6}{8}$
$+ 2\frac{4}{8}$
$\overline{5\frac{10}{8}}$

Step 3:
Rename improper fractions and reduce to simplest form.

$5\frac{10}{8} = 5 + 1\frac{2}{8} = 6\frac{2}{8} = 6\frac{1}{4}$

So, $3\frac{6}{8} + 2\frac{1}{2} = 6\frac{1}{4}$.

Directions

Add. Reduce to simplest form.

1. $3\frac{5}{8}$
$+ 2\frac{1}{8}$

2. $6\frac{1}{3}$
$+ 2\frac{7}{12}$

3. $4\frac{1}{4}$
$+ 2\frac{1}{4}$

4. $5\frac{5}{7}$
$+ 1\frac{5}{14}$

5. $1\frac{5}{9}$
$+ 2\frac{8}{9}$

6. $3\frac{3}{4}$
$+ \quad \frac{4}{8}$

7. $3\frac{3}{10}$
$+ 2\frac{1}{2}$

8. $2\frac{1}{6}$
$+ 3\frac{7}{12}$

Name _____   Date _____

Subtracting Fractions with Unlike Denominators

When subtracting fractions, remember to look at the denominators first. If they are not the same, find a **common denominator**.

Example: Subtract $\frac{3}{4} - \frac{3}{5} = \boxed{}$

Step 1: Find a common denominator. Choose the least common multiple.

multiples of 4: 4, 8, 12, 16, ⟨20,⟩ 24, ...
multiples of 5: 5, 10, 15, ⟨20,⟩ 25, ...
LCM: 20

Step 2: Rename each fraction using the common denominator.

$\frac{3}{4} = \frac{15}{20}$ $\frac{3}{5} = \frac{12}{20}$

Step 3: Subtract the fractions.

$\frac{15}{20} - \frac{12}{20} = \frac{3}{20}$

Step 4: Write the difference in simplest form.

$\frac{3}{20}$

So, $\frac{3}{4} - \frac{3}{5} = \frac{3}{20}$.

Directions

Subtract. Reduce to simplest form.

1. $\frac{4}{6}$
$-\frac{1}{3}$

2. $\frac{3}{4}$
$-\frac{4}{6}$

3. $\frac{9}{14}$
$-\frac{3}{7}$

4. $\frac{5}{6}$
$-\frac{5}{8}$

5. $\frac{3}{9}$
$-\frac{1}{3}$

6. $\frac{4}{5}$
$-\frac{3}{10}$

7. $\frac{11}{12}$
$-\frac{3}{4}$

8. $\frac{13}{16}$
$-\frac{3}{8}$

9. $\frac{1}{2}$
$-\frac{1}{12}$

10. $\frac{2}{3}$
$-\frac{4}{9}$

Subtracting Mixed Numbers

When subtracting mixed numbers, first subtract the fractions, then subtract the whole numbers. Sometimes, you will need to rename fractions before you subtract.

Example: Subtract $2\frac{1}{3} - 1\frac{4}{6} = \boxed{}$

Step 1:
Rename each fraction using a common denominator.

$$2\frac{1}{3} \longrightarrow 2\frac{2}{6}$$
$$-1\frac{4}{6} \longrightarrow 1\frac{4}{6}$$

Step 2:
Rename the minuend as an improper fraction.

$$2\frac{2}{6} \longrightarrow 1\frac{8}{6}$$
$$-1\frac{4}{6} \longrightarrow 1\frac{4}{6}$$

Step 3:
Subtract the fractions. Subtract the whole numbers.

$$1\frac{8}{6}$$
$$-1\frac{4}{6}$$
$$\overline{\frac{4}{6}}$$

Step 4:
Reduce to simplest form.

$$\frac{4}{6} = \frac{2}{3}$$

So, $2\frac{1}{3} - 1\frac{4}{6} = \frac{2}{3}$.

Directions

Subtract. Reduce to simplest form.

1. $3\frac{1}{5}$
    $-1\frac{2}{5}$

2. $4\frac{2}{4}$
    $-3\frac{1}{4}$

3. $7\frac{4}{9}$
    $-2\frac{1}{3}$

4. $5\frac{1}{2}$
    $-2\frac{5}{6}$

5. $5\frac{4}{7}$
    $-3\frac{2}{14}$

6. $6\frac{1}{4}$
    $-3\frac{7}{8}$

7. $2\frac{1}{2}$
    $-1\frac{3}{4}$

8. $6\frac{2}{3}$
    $-1\frac{1}{6}$

9. 4
    $-\frac{2}{5}$

10. $9\frac{1}{9}$
    -3

Name _____ Date _____

Multiplying a Fraction by a Whole Number

When you multiply a fraction by a whole number, remember to rename the whole number as a fraction.

Multiply $\frac{2}{5} \times 10 = \boxed{}$

Example 1: Use models.

Step 1: Use fraction circles to show the whole number.

Step 2: Show the shaded fraction in each circle.

$\frac{20}{5}$ in all.

Step 3: Rename the improper fraction. $\frac{20}{5} = 4$

Example 2: Use calculating.

Step 1: Rename the whole number as a fraction. $10 = \frac{10}{1}$

Step 2: Multiply the numerators. Multiply the denominators. $\frac{2}{5} \times \frac{10}{1} = \frac{2 \times 10}{5 \times 1} = \frac{20}{5}$

Step 3: Reduce to simplest form. $\frac{20}{5} = 4$

So, $\frac{2}{5} \times 10 = 4$.

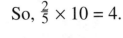

Multiply. Use models or calculating.

1. $\frac{1}{6} \times 18 =$ _____

2. $\frac{1}{7} \times 21 =$ _____

3. $\frac{1}{4} \times 16 =$ _____

4. $\frac{3}{8} \times 24 =$ _____

5. $\frac{2}{7} \times 14 =$ _____

6. $\frac{5}{8} \times 24 =$ _____

Multiplying a Fraction by a Fraction

When you multiply two fractions, each of which is less than 1, the product is smaller than either of the two fractions.

Multiply $\frac{1}{2} \times \frac{2}{3} = \boxed{}$, or $\frac{1}{2}$ of $\frac{2}{3}$

Example 1: Use models.

Step 1: Use a model to show $\frac{2}{3}$.

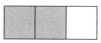

Step 2: Shade $\frac{1}{2}$ of the shading in the above model.

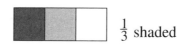 $\frac{1}{3}$ shaded

Example 2: Use calculating.

Step 1: Multiply the numerators. Multiply the denominators.

$$\frac{1}{2} \times \frac{2}{3} = \frac{1 \times 2}{2 \times 3} = \frac{2}{6}$$

Step 2: Reduce to simplest form.

$$\frac{2}{6} = \frac{1}{3}$$

So, $\frac{1}{2} \times \frac{2}{3} = \frac{1}{3}$.

Directions

Use models or calculating. Reduce to simplest form.

1. $\frac{1}{4}$ of $\frac{1}{2}$ = _____

2. $\frac{1}{3}$ of $\frac{1}{2}$ = _____

3. $\frac{2}{3}$ of $\frac{1}{4}$ = _____

4. $\frac{1}{3}$ of $\frac{3}{5}$ = _____

5. $\frac{1}{2} \times \frac{4}{8}$ = _____

6. $\frac{1}{6} \times \frac{3}{4}$ = _____

7. $\frac{3}{8} \times \frac{2}{3}$ = _____

8. $\frac{4}{5} \times \frac{2}{5}$ = _____

Name _____ Date _____

Multiplying a Fraction by a Mixed Number

When you multiply a fraction by a mixed number, you can use the Distributive Property to break apart numbers to multiply.

Multiply $\frac{1}{2} \times 1\frac{3}{4} = \boxed{}$

Example 1: Use models.

   Step 1: Use a model to show $1\frac{3}{4}$.

   Step 2: Shade $\frac{1}{2}$ of each figure.

$\frac{4}{8}$ shaded $\frac{3}{8}$ shaded

   Step 3: Add the shaded parts.

$\frac{4}{8} + \frac{3}{8} = \frac{7}{8}$

Example 2: Use calculating.

   Step 1: Break apart the mixed number. $\frac{1}{2} \times (1 + \frac{3}{4})$

   Step 2: Multiply each part. $(\frac{1}{2} \times 1) + (\frac{1}{2} \times \frac{3}{4})$

   Step 3: Add the products. $\frac{1}{2} + \frac{3}{8}$

   Step 4: Find the LCD and rename the fractions. $\frac{4}{8} + \frac{3}{8}$

   Step 5: Reduce to simplest form if you can. $\frac{7}{8}$

So, $\frac{1}{2} \times 1\frac{3}{4} = \frac{7}{8}$.

Directions

Use models or calculating. Reduce to simplest form.

1. $\frac{1}{3} \times 3\frac{1}{5} =$ _____

2. $\frac{1}{2} \times 2\frac{3}{4} =$ _____

3. $\frac{1}{6} \times 3\frac{2}{3} =$ _____

4. $\frac{1}{4} \times 2\frac{5}{6} =$ _____

5. $\frac{1}{3} \times 3\frac{1}{2} =$ _____

6. $\frac{1}{8} \times 4\frac{1}{4} =$ _____

Dividing Fractions

When you divide fractions, you use the **reciprocal number** of the divisor. The **reciprocal** is made by changing the places of the numerator and denominator. The product of a number and its reciprocal is 1. The reciprocal of $\frac{2}{3}$ is $\frac{3}{2}$. Then, write a multiplication sentence and solve.

Example: Divide $\frac{1}{2} \div \frac{1}{4} = \boxed{}$

Step 1: Write the reciprocal of the divisor.   The reciprocal of $\frac{1}{4}$ is $\frac{4}{1}$.

Step 2: Use the reciprocal to write a multiplication sentence.

$$\frac{1}{2} \times \frac{4}{1} = \boxed{}$$

Step 3: Multiply.

$$\frac{1}{2} \times \frac{4}{1} = \frac{1 \times 4}{2 \times 1} = \frac{4}{2}$$

Step 4: Reduce to simplest form.

$$\frac{4}{2} = 2$$

So, $\frac{1}{2} \div \frac{1}{4} = 2$.

Directions

Write the reciprocal number.

1. $\frac{1}{2}$ _____

2. $\frac{2}{3}$ _____

3. $\frac{3}{8}$ _____

4. $\frac{5}{2}$ _____

5. 4 _____

6. 9 _____

Directions

Divide. Reduce to simplest form.

7. $\frac{1}{3} \div \frac{2}{3} =$ _____

8. $\frac{2}{4} \div \frac{1}{8} =$ _____

9. $\frac{2}{6} \div \frac{1}{4} =$ _____

10. $\frac{2}{5} \div \frac{2}{5} =$ _____

11. $4 \div \frac{1}{2} =$ _____

12. $\frac{3}{4} \div 2 =$ _____

Name _____   Date _____

Unit 3 Assessment

Directions

Solve.

1. Alan uses $\frac{1}{4}$ cup of walnuts to make nutbread and $\frac{3}{8}$ cup to make the pineapple delight. How many cups of walnuts does he use in all?

_____

2. Jason swam in 8 out of 9 meets. Becky swam in 4 out of 9 meets. Who swam in about one-half of the meets?

_____

3. 9 cups = _____ pints

4. 5 quarts = _____ gallons

Directions

Write or draw what comes next.

5. $\frac{1}{4}, \frac{2}{6}, \frac{3}{8},$ _____

6.  _____

Directions

Use the graph to answer the questions.

Mr. Alholm works in a men's shoe store. The graph shows the number of shoes he sold on Tuesday.

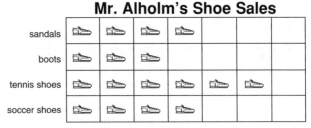

Mr. Alholm's Shoe Sales

sandals	👟	👟	👟	👟		
boots	👟	👟	👟			
tennis shoes	👟	👟	👟	👟	👟	👟
soccer shoes	👟	👟	👟	👟		

👟 = 2 shoes (1 pair)

7. How many pairs of shoes did Mr. Alholm sell in all? _____

8. What fraction of his sales was tennis shoes? _____

9. What fraction of his sales was boots? _____

Word Problems

Directions

Solve.

1. Petra loves animals. She has 12 pets in all, 4 of which are rabbits. Write a fraction to describe the number of rabbits she has.

   _____

2. Gilbert is making snack mix for a school field trip. He uses $\frac{1}{2}$ cup dried apricots, $\frac{3}{5}$ cup peanuts, and $\frac{4}{7}$ cup raisins. List the ingredients in order from least to greatest.

   _____

3. Phil biked $\frac{2}{8}$ mile on Monday and $\frac{5}{8}$ mile on Tuesday. On which day did he bike farther? How much farther did he bike?

   _____

4. Joanna talked on the phone for $\frac{3}{5}$ hour before dinner. She talked on the phone for $\frac{11}{12}$ hour after dinner. About how long did she talk in all?

   _____

5. Martha spent $2\frac{1}{2}$ hours reading on Saturday. She spent $\frac{3}{4}$ of an hour reading on Sunday. How many hours did she spend reading this weekend?

   _____

6. Martha had $3\frac{1}{3}$ dozen eggs. She used $\frac{3}{4}$ dozen to make cookies for a bake sale. How many dozen eggs does she have left?

   _____

7. Joan is making valentines. She has $\frac{2}{3}$ sheet of paper. Each valentine takes $\frac{1}{6}$ sheet. How many valentines can Joan make?

   _____

8. Jade swam $\frac{1}{2}$ mile on Monday. On Wednesday she swam $\frac{3}{8}$ mile. How many miles did Jade swim in all?

   _____

Patterns with Fractions

Directions

Answer the questions.

1.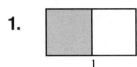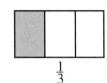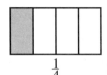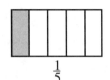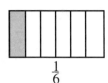

   $\frac{1}{2}$ $\frac{1}{3}$ $\frac{1}{4}$ $\frac{1}{5}$ $\frac{1}{6}$

   a. What is the pattern?

   _____

   b. What do you notice about the shaded parts?

   _____

2.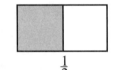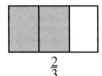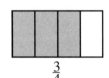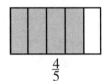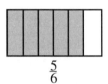

   $\frac{1}{2}$ $\frac{2}{3}$ $\frac{3}{4}$ $\frac{4}{5}$ $\frac{5}{6}$

   a. What is the pattern?

   _____

   b. What do you notice about the shaded parts?

   _____

Directions

Complete the pattern.

3.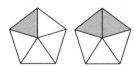

4.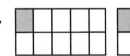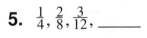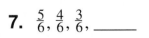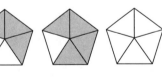

5. $\frac{1}{4}, \frac{2}{8}, \frac{3}{12},$ _____

6. $\frac{1}{2}, 1, 1\frac{1}{2}, 2,$ _____

7. $\frac{5}{6}, \frac{4}{6}, \frac{3}{6},$ _____

8. $\frac{1}{2}, \frac{2}{5}, \frac{3}{8}, \frac{4}{11},$ _____

Name _____ Date _____

More Patterns with Fractions

Some problems are solved by looking for a pattern. You can use a sequence to find the rule to continue the pattern.

Example: Tia is training to run a race. During the first week of training, she runs $\frac{1}{3}$ mi each day. During the second week she runs $\frac{1}{2}$ mi each day. During the third week, she runs $\frac{2}{3}$ mi each day. During which week will she run $1\frac{1}{2}$ mi each day?

Step 1: Write a sequence to show the miles Tia ran for the first three weeks.

$$\frac{1}{3}, \frac{1}{2}, \frac{2}{3}, \dots$$

Step 2: Find the LCD for the fractions. The LCD of 2 and 3 is 6. Use the LCD to write fractions with the same denominator.

$$\frac{2}{6}, \frac{3}{6}, \frac{4}{6}, \dots$$

Step 3: Look for a pattern. Add $\frac{1}{6}$ to each term to find the next term.

Step 4: Continue the pattern to find the week Tia will run $1\frac{1}{2}$ mi.

$$\frac{2}{6}, \frac{3}{6}, \frac{4}{6}, \frac{5}{6}, 1, 1\frac{1}{6}, 1\frac{2}{6}, 1\frac{3}{6}$$

So, Tia will run $1\frac{1}{2}$ mi each day during the eighth week.

Directions

Solve.

1. Mark is reading a book. On July 5, he read for $\frac{1}{2}$ hr. On July 6 he read for $\frac{7}{10}$ hr. On July 7 he read for $\frac{9}{10}$ hr. If he continues reading in this manner, how long will he read on July 12?

   _____

2. Keiko is painting a fence. She painted for $\frac{3}{4}$ hr on the first day, $1\frac{1}{2}$ hr the second day, and $2\frac{1}{4}$ hr the third day. If she continues this pattern, how much time will she spend painting on the fourth day?

   _____

Name _____   Date _____

Time

Directions

Write the elapsed time as a fraction.

1.

Begin   End

A.M.   P.M.

2.

Begin   End

P.M.   A.M.

Elapsed time: _____   Elapsed time: _____

Directions

Use the table to solve the problems.

Ms. Jacks gave a lot of homework during the weekend. The table shows how much of the day some of her students spent on it.

3. How much more of the day did Mark spend on homework than Theresa?

_____

4. Who spent the greatest fraction of the day working? the least? What is the difference between the two fractions?

_____

_____

5. How many days' worth of time did Lisa, Pete, and Chris spend in all on homework?

_____

Time Spent on Homework

Name	Time
Mark	$\frac{3}{4}$ of the day
Lisa	$\frac{3}{8}$ of the day
Pete	$\frac{1}{8}$ of the day
Chris	$\frac{5}{8}$ of the day
Theresa	$\frac{1}{2}$ of the day
Charlie	$\frac{1}{4}$ of the day

Name _____ Date _____

Capacity

Customary Units of Capacity

| 8 fluid ounces (fl oz) = 1 cup (c) |
| 2 c = 1 pint (pt) |
| 2 pt = 1 quart (qt) |
| 4 qt = 1 gallon (gal) |

Directions

Write the fraction.

1. 5 cups = _____ pints

2. 3 pints = _____ quarts

3. 12 pints = _____ gallons

4. 7 cups = _____ quarts

5. 18 fluid ounces = _____ cups

6. 9 quarts = _____ gallons

Directions

Find the amount needed of each ingredient to cut the recipe in half.

Waffles

2 eggs	$\frac{1}{2}$ teaspoon baking soda	$\frac{3}{4}$ cup milk
$\frac{7}{8}$ cup flour	$\frac{1}{4}$ teaspoon salt	$\frac{2}{3}$ cup butter
2 teaspoons baking powder	2 tablespoons sugar	$\frac{1}{2}$ teaspoon vanilla
	$\frac{3}{4}$ cup sour cream	

7. eggs   **8.** flour   **9.** baking powder   **10.** baking soda

_____   _____   _____   _____

11. salt   **12.** sugar   **13.** sour cream   **14.** milk

_____   _____   _____   _____

15. butter   **16.** vanilla

_____   _____

Measurement

Customary Units of Length

| 12 inches (in.) = 1 foot (ft) |
| 3 feet (ft) = 1 yard (yd) |
| 5,280 ft or 1,760 yd = 1 mile (mi) |

Directions **Measure the width of your desk to the nearest inch, $\frac{1}{2}$ inch, $\frac{1}{4}$ inch, $\frac{1}{8}$ inch, and $\frac{1}{16}$ inch.**

1. Which measurement is the least precise? _____

2. Which measurement is the most precise? _____

Directions **Measure each line segment to the part of an inch that gives the most precise measurement.**

3. ━━━━━━━━━━━━━━━━━━━━━━━━━━   _____

4. ━━━━━━━━━━━━━━━   _____

5. ━━━━━━━━━━━━━━━━━━━━   _____

Directions **Draw a line to the given length.**

6. $4\frac{1}{2}$ in.

7. $3\frac{1}{8}$ in.

8. $1\frac{1}{2}$ in.

9. Sam made the chart at the right to keep track of how much wood he had for projects. He forgot to enter some of the numbers. Complete the table.

10. Each week Henry will work $3\frac{1}{2}$ hours on Wednesday and $4\frac{1}{4}$ hours on Friday. How many hours will he work each week?

Type of Wood	Start With	Feet Used	Feet Left
Oak	$15\frac{1}{2}$	$9\frac{1}{4}$	
Pine	$22\frac{5}{8}$		$10\frac{1}{4}$
Maple		$12\frac{3}{4}$	$2\frac{1}{6}$
Cherry	$20\frac{3}{4}$	$5\frac{3}{8}$	

_____

Probability

> **Probability** is the chance that an activity or event will happen. The outcome is the result of the activity or event. Probability can be shown as a fraction.

Example: Ben has a spinner with six sections. The possible outcomes are spinning blue, spinning red, spinning yellow, or spinning green. What is the probability of spinning blue?

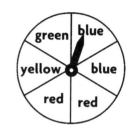

Probability of spinning blue = $\dfrac{\text{number of blue sections}}{\text{total number of sections}} = \dfrac{2}{6}$

So, the probability of spinning blue is $\frac{2}{6}$, or $\frac{1}{3}$.

Directions

For Problems 1–4, use spinner A. Give the probability of spinning each color.

1. blue _____

2. red _____

3. green _____

4. yellow _____

A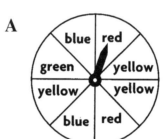

Directions

For Problems 5–9, use spinner B. Give the probability of spinning each number.

5. 1 _____

6. 2 _____

7. 3 _____

8. 4 _____

9. 5 _____

B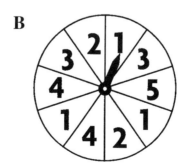

Name _____   Date _____

Bar Graphs

A **bar graph** uses bars to show information. The bars can be vertical or horizontal. The heights and lengths of the bars make it easier to compare data.

The bar graph shows the time of year students in
Mr. Monroe's class have a birthday.

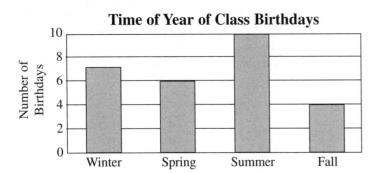

Directions

Use the graph to answer the questions.

1. In what season did the most students have birthdays? _____

2. How many students had birthdays in the winter? _____

3. How many students are in Mr. Monroe's class? _____

4. What fraction would you write to show the number of months in winter? _____

5. Write the fraction of students who have birthdays in each season.

   a. winter _____

   b. spring _____

   c. summer _____

   d. fall _____

Fractions 5, SV 3408-8

Line Graphs

A **line graph** shows how data change over time. The scale runs along the left side of the graph. The time period is set along the bottom of the graph. If the points of the line graph are between numbers of the scale, you will need to estimate.

The line graph shows the bicycles sold in one week at a sports store.

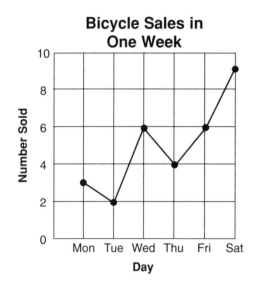

Directions

Use the graph to answer the questions.

1. On what day were the least number of bicycles sold? _____

2. On what days were the same number of bicycles sold? _____

3. How many bicycles were sold in all? _____

4. Write in simplest form the fraction of bicycles sold on each day.

   a. Wednesday _____   **b.** Tuesday _____

   c. Monday _____   **d.** Saturday _____

Name _____   Date _____

Circle Graphs

A **circle graph** looks like it sounds. Each slice of the circle is a piece of
information included in the circle. The size of each piece shows its relationship
to the whole and to each piece in the circle. A circle graph can use fractions to
show data. The sum of the fractional pieces must equal 1.

The circle graph shows the fractional
amount of ingredients used to make punch.

Punch for a Crowd

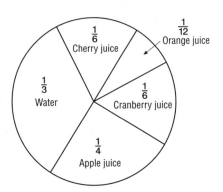

(**Directions**)

Use the graph to answer the questions.

1. What fraction of apple juice is used? _____

2. What fraction of juice is used in all? _____

3. Arrange the ingredients in order from least to greatest.

_____

_____

4. If a caterer makes 12 gallons of punch, how much of each ingredient is needed?

   a. water _____   **b.** cherry juice _____

   c. orange juice _____   **d.** cranberry juice _____

   e. apple juice _____

Geometry

Directions

Write the fraction or mixed number of shaded parts in simplest form.

1.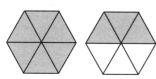

2.

_____   _____

Directions

Write a mixed number in simplest form for each of the following figures. The figure at the right stands for 1.

3.

4.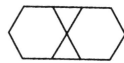

_____   _____

5.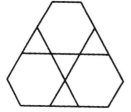

6.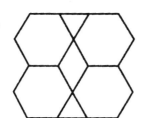

7.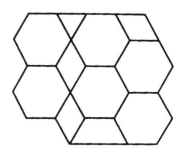

_____   _____   _____

Directions

Shade parts of the following figures. Have a partner write a mixed number that tells how much is shaded.

8.

9.

_____   _____

Algebra

> A letter can take the place of a number in a number sentence. The letter represents the number for which you solve.

Example 1: Solving an addition problem in which an addend is unknown.

Step 1: Look for a way to get x alone on one side of the equation. Use the inverse operation on both sides of the equation.

$$x + \tfrac{1}{2} = \tfrac{3}{4}$$
$$x + \tfrac{1}{2} - \tfrac{1}{2} = \tfrac{3}{4} - \tfrac{1}{2}$$
$$x + 0 = \tfrac{3}{4} - \tfrac{1}{2}$$
$$x = \tfrac{3}{4} - \tfrac{1}{2}$$

Step 2: Find the LCD.

$$x = \tfrac{3}{4} - \tfrac{2}{4}$$

Step 3: Subtract.

$$x = \tfrac{1}{4}$$

So, $\tfrac{1}{4} + \tfrac{1}{2} = \tfrac{3}{4}$.

Example 2: Solving an subtraction problem in which the minuend is unknown.

Step 1: Look for a way to get x alone on one side of the equation. Use the inverse operation on both sides of the equation.

$$x - \tfrac{2}{3} = \tfrac{1}{6}$$
$$x - \tfrac{2}{3} + \tfrac{2}{3} = \tfrac{1}{6} + \tfrac{2}{3}$$
$$x - 0 = \tfrac{1}{6} + \tfrac{2}{3}$$
$$x = \tfrac{1}{6} + \tfrac{2}{3}$$

Step 2: Find the LCD.

$$x = \tfrac{1}{6} + \tfrac{4}{6}$$

Step 3: Add.

$$x = \tfrac{5}{6}$$

So, $\tfrac{5}{6} - \tfrac{2}{3} = \tfrac{1}{6}$.

Directions

Solve. Reduce to simplest form.

1. $x + \tfrac{2}{3} = \tfrac{5}{6}$ $x =$ _____

2. $x - \tfrac{1}{2} = \tfrac{1}{4}$ $x =$ _____

3. $x - \tfrac{1}{6} = \tfrac{2}{3}$ $x =$ _____

4. $x + \tfrac{5}{12} = \tfrac{7}{12}$ $x =$ _____

5. $x + \tfrac{1}{3} = \tfrac{8}{9}$ $x =$ _____

6. $x - \tfrac{3}{8} = \tfrac{1}{2}$ $x =$ _____

Equivalent Fraction Patterns

Each series of equivalent fractions creates a pattern. One fraction in each series does not follow the pattern. Cross out that fraction. Write the fraction that belongs in its place. The first one is done for you.

1. $\frac{2}{3}$ | ~~$\frac{6}{7}$~~ | $\frac{6}{9}$ | $\frac{8}{12}$ | $\underline{\quad \frac{4}{6} \quad}$

2. $\frac{4}{16}$ | $\frac{2}{12}$ | $\frac{2}{8}$ | $\frac{1}{4}$ | _____

3. $\frac{2}{16}$ | $\frac{3}{24}$ | $\frac{4}{30}$ | $\frac{5}{40}$ | _____

4. $\frac{2}{4}$ | $\frac{4}{8}$ | $\frac{6}{16}$ | $\frac{16}{32}$ | _____

5. $\frac{3}{4}$ | $\frac{6}{8}$ | $\frac{10}{12}$ | $\frac{12}{16}$ | _____

6. $\frac{2}{7}$ | $\frac{4}{14}$ | $\frac{8}{21}$ | $\frac{16}{56}$ | _____

7. $\frac{25}{28}$ | $\frac{15}{21}$ | $\frac{10}{14}$ | $\frac{5}{7}$ | _____

8. $\frac{1}{4}$ | $\frac{3}{10}$ | $\frac{9}{36}$ | $\frac{27}{108}$ | _____

9. $\frac{20}{32}$ | $\frac{15}{24}$ | $\frac{10}{16}$ | $\frac{6}{8}$ | _____

10. $\frac{9}{11}$ | $\frac{18}{22}$ | $\frac{24}{33}$ | $\frac{36}{44}$ | _____

11. $\frac{4}{10}$ | $\frac{8}{20}$ | $\frac{10}{30}$ | $\frac{16}{40}$ | _____

12. $\frac{1}{5}$ | $\frac{2}{15}$ | $\frac{3}{15}$ | $\frac{4}{20}$ | _____

Decimals and Fractions

A **decimal** is a number that uses place value and a decimal to show a value less than 1. A fraction can name a decimal.

Model	Fraction	Decimal Value			Decimal	Read
	3 shaded parts / 10 parts	O 0	. T 3	H	0.3	three tenths
	64 shaded parts / 100 parts	O 0	. T 6	H 4	0.64	sixty-four hundredths

Directions

Write the fraction and decimal.

1.

fraction _____

decimal _____

2.

fraction _____

decimal _____

Directions

Write the fraction or decimal.

3. 1.32 _____

4. $3\frac{2}{10}$ _____

5. 3.05 _____

6. 0.70 _____

7. $\frac{54}{100}$ _____

8. $\frac{4}{100}$ _____

Directions

Round to the underlined digit.

9. 5.7̲8 _____

10. 1̲.62 _____

11. 0.2̲4 _____

Name _____ Date _____

Using a Calculator

You can "fool" your calculator into accepting fractions and mixed numbers. For example, you can use this key sequence to input $3\frac{5}{8}$.

Press: 5 [÷] 8 [+] 3 [=] 3.625

The result is a decimal that is equivalent to the mixed number.

1. Press 3 + 5 ÷ 8. Is the result the same as the one in the example? Explain why or why not.

   _____

   _____

2. Explain why you should enter the fraction before you add the whole number.

   _____

   _____

Directions

Use your calculator to find the decimal equivalent to the mixed number. If necessary, round your answer to the nearest hundredth.

3. $4\frac{3}{5}$ _____

4. $7\frac{9}{11}$ _____

5. $8\frac{3}{7}$ _____

6. $5\frac{7}{8}$ _____

You can also "fool" your calculator into adding fractions and mixed numbers. For example, find $3\frac{2}{5} + 2\frac{1}{4}$.

Press: 2 [÷] 5 [+] 3 [M+] [C] 1 [÷] 4 [+] 2 [M+] [MRC] [=] 5.65

Directions

Use your calculator to find the decimal equivalent to the sum. If necessary, round your answer to the nearest hundredth.

7. $4\frac{1}{2} + 7\frac{4}{5} =$ _____

8. $2\frac{4}{7} + 6\frac{3}{5} =$ _____

Name _____ Date _____

Ratios

Ratios compare two numbers. There are three types of ratios. Each ratio can be written in three ways.

Part to Whole	Whole to Part	Part to Part
	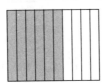	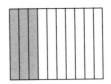

Part to Whole

Shaded parts: 4
Total parts: 10
So, the ratio of part to whole is 4 to 10.
It can be written 4 to 10, 4:10, or $\frac{4}{10}$.

Whole to Part

Total parts: 10
Shaded parts: 6
So, the ratio of whole to part is 10 to 6.
It can be written 10 to 6, 10:6, or $\frac{10}{6}$.

Part to Part

Shaded parts: 3
Unshaded parts: 7
So, the ratio of part to part is 3 to 7.
It can be written 3 to 7, 3:7, or $\frac{3}{7}$.

Directions

Find the ratios.

1.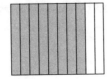

    a. shaded parts: _____

    b. total parts: _____

    c. part to whole ratio: _____

2.

    a. shaded parts: _____

    b. unshaded parts: _____

    c. part to part ratio: _____

Directions

Circle the letter of the *incorrect* ratio.

3. 3 red apples out of 8 apples
    a. $\frac{3}{8}$ **b.** 8:3   **c.** 3 to 8

4. 7 boys to 8 girls
    a. $\frac{8}{7}$ **b.** 7:8   **c.** 7 to 8

5. 8 baseballs and 13 basketballs
    a. $\frac{8}{13}$ **b.** 8:13   **c.** 13 to 8

6. 1 month out of 12 months
    a. $\frac{12}{1}$ **b.** 1:12   **c.** 1 to 12

Unit 4: Enrichment

Fractions 5, SV 3408-8

Percents

A **percent** is a ratio in which the denominator is 100. 45% means forty-five hundredths, or $\frac{45}{100}$.

Example 1: Write a percent as a fraction. 50%

    Step 1: Write the percent as a fraction. $50\% = \frac{50}{100}$

    Step 2: Write the fraction in simplest form. $\frac{50}{100} = \frac{50 \div 50}{100 \div 50} = \frac{1}{2}$

So, $50\% = \frac{1}{2}$.

Example 2: Write a fraction as a percent. $\frac{3}{4}$

    Step 1: Write the fraction with 100 as the denominator. $\frac{3}{4} = \frac{3 \times 25}{4 \times 25} = \frac{75}{100}$

    Step 2: Write the fraction as a percent. $\frac{75}{100} = 75\%$

So, $\frac{3}{4} = 75\%$.

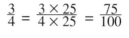

Directions

Write the percent as a fraction in simplest form.

  1. 50% _____   **2.** 30% _____   **3.** 25% _____

  4. fifty-six percent _____   **5.** twenty-three percent _____

Directions

Write the fraction as a percent.

  6. $\frac{19}{100}$ _____   **7.** $\frac{8}{25}$ _____   **8.** $\frac{71}{100}$ _____

  9. $\frac{2}{100}$ _____   **10.** $\frac{2}{10}$ _____   **11.** $\frac{1}{4}$ _____

Money

The value of money can be written as a fraction or as a decimal.	100 pennies = 1 dollar
	20 nickels = 1 dollar
	10 dimes = 1 dollar
	4 quarters = 1 dollar

Directions

Write a fraction and a decimal to tell what part of each coin makes a dollar. Write the fractions in simplest form.

1. a penny

_____

2. a nickel

_____

3. a dime

_____

4. a quarter

_____

Directions

Write each amount as a fraction and as a decimal.

5.  _____

_____

6.  _____

_____

| The hobby shop is having a sale on stamp albums. |
| $\frac{1}{3}$ of $15 = $5   You save $5. |

Directions

Write the amount saved.

7. Regular Price: $10

$\frac{1}{4}$ off _____

8. Regular Price: $20

$\frac{1}{3}$ off _____

Fractions: Grade 5

Answer Key

p. 5
1. B
2. D
3. D
4. C
5. D
6. B
7. A
8. B

p. 6
9. A
10. B
11. D
12. D
13. C
14. C
15. A
16. C

p. 7
1. 3/4
2. 2/5
3. <
4. >
5. <
6. 2/5
7. 3/8
8. 1/2
9. 2
10. 3
11. 4
12. 1/4, 1/3, 7/8
13. 2 1/9, 2 3/5, 4 1/9

p. 8
1. 3/4
2. 2/4
3. 3/6
4. Students shade 2 parts.
5. Students shade 1 triangle.
6. Students shade 5 squares.

p. 9
1. about 1/2
2. about 0
3. about 1
4. 5/8
5. 2/3
6. 3/4

p. 10
1. yes
2. no
3. 6/16
4. 10/15
5. 3/4
6. 1/4

p. 11
1. 20
2. 24
3. 10
4. 3: 3, 6, 9, 12, 15
   4: 4, 8, 12, 16, 20, 24
   LCM: 12
5. 3: 3, 6, 9, 12, 15
   5: 5, 10, 15, 20, 25
   LCM: 15
6. 3: 3, 6, 9, 12, 15
   6: 6, 12, 18, 24, 30
   LCM: 6

p. 12
1. 3
2. 3; 2
3. $3 \times 2 = 6$; 6
4. 4
5. 3
6. 8

p. 13
1. GCF: 2; 3/4
2. $14 = 7 \times 2$; $21 = 7 \times 3$;
   GCF: 7; 2/3
3. 3/5
4. 5/7
5. 2/3

p. 14
1. 2/4
2. 2/3
3. 7/8
4. 1/4
5. 1/2
6. 3/5
7. >
8. <
9. =
10. =
11. <
12. >

p. 15
1. 1/4, 1/3, 1/2
2. 1/6, 1/4, 1/3
3. 1/10, 1/5, 1/4
4. 1/4, 3/8, 1/2
5. 1/3, 4/9, 5/6
6. 7/12, 2/3, 3/4

p. 16
1. 1 2/3
2. 2 3/8
3. 1 1/4
4. 1 4/5
5. 1 4/6 or 1 2/3

p. 17
1. 1 1/6
2. 3/7
3. 9/10
4. 1 1/3
5. 13/14
6. 6 7/10
7. 3 3/8
8. 3 1/3
9. 1
10. 1/6
11. 1 2/3
12. 2/9
13. 9/10
14. 2/3
15. 1/8

p. 18
1. 1
2. 0
3. 1/2
4. 0
5. 1/2
6. 1
7. 1
8. 1
9. 0
10. 1/2
11. 1/2
12. 0

p. 19
1. $1 + 2$; 3; 1/2
2. $7 - 4$; 3; 1/3
3. 1/2
4. 1/5
5. 1
6. 2/3
7. 2/9
8. 1 1/3
9. 1/4
10. 1 1/2

p. 20
1. 6
2. 4
3. 10
4. 12
5. 7/10
6. 1 5/12
7. 7/12
8. 1 1/3
9. 9/10
10. 1 1/12
11. 1 1/8
12. 5/6

p. 21
1. 5 3/4
2. 8 11/12
3. 6 1/2
4. 7 1/14
5. 4 4/9
6. 4 1/4
7. 5 4/5
8. 5 3/4

p. 22
1. 1/3
2. 1/12
3. 3/14
4. 5/24
5. 0
6. 1/2
7. 1/6
8. 7/16
9. 5/12
10. 2/9

p. 23
1. 1 4/5
2. 1 1/4
3. 5 1/9
4. 2 2/3
5. 2 3/7
6. 2 3/8
7. 3/4
8. 5 1/2
9. 3 3/5
10. 6 1/9

p. 24
1. 3
2. 3
3. 4
4. 9
5. 4
6. 15

p. 25
1. 1/8
2. 1/6
3. 1/6
4. 1/5
5. 1/4
6. 1/8
7. 1/4
8. 8/25

p. 26
1. 1 1/15
2. 1 3/8
3. 11/18
4. 17/24
5. 1 1/6
6. 17/32

p. 27
1. 2
2. 3/2
3. 8/3
4. 2/5
5. 1/4
6. 1/9
7. 1/2
8. 4
9. 1 1/3
10. 1
11. 8
12. 3/8

Fractions: Grade 5

Answer Key (cont.)

p. 28
1. 5/8 cup
2. Becky
3. 4 1/2
4. 1 1/4
5. 4/10
6.
7. 17
8. 6/17
9. 3/17

p. 29
1. 1/3 rabbits
2. apricots, raisins, peanuts
3. Tuesday; 3/8 mi
4. about 1 1/2 hr
5. 3 1/4 hr
6. 2 7/12 dozens
7. 4 valentines
8. 7/8 mi

p. 30
1. a. The denominator increases by 1.
   b. The parts get smaller as the denominator gets larger.
2. a. The numerator and the denominator increase by 1.
   b. The shaded part gets larger as both numbers get larger.
3. Students shade all 5 parts.
4. Students repeat the pattern plus shade the fourth square on the bottom row.
5. 4/16
6. 2 1/2
7. 2/6
8. 5/14

p. 31
1. 1/2, 7/10, 9/10, ...;
   1 9/10 hr
2. 3/4, 1 1/2, 2 1/4, ...; 3 hr

p. 32
1. 9 1/2 hr
2. 4 1/6 hr
3. 1/4 more
4. greatest-Mark;
   least-Pete; difference-5/8
5. 1 1/8 days

p. 33
1. 2 1/2
2. 1 1/2
3. 1 1/2
4. 1 3/4
5. 2 1/4
6. 2 1/4
7. 1 egg
8. 7/16 c
9. 1 tsp
10. 1/4 tsp
11. 1/8 tsp
12. 1 tbsp
13. 3/8 c
14. 3/8 c
15. 1/3 c
16. 1/4 tsp

p. 34
1. inch
2. 1/16 in.
3. 4 1/2 in.
4. 2 3/4 in.
5. 3 3/8 in.
For 6–8, check students' lines.
9. Oak-6 1/4
   Pine-12 3/8
   Maple-14 11/12
   Cherry-15 3/8
10. 7 3/4 hr

p. 35
1. 1/4
2. 1/4
3. 1/8
4. 3/8
5. 3/10
6. 1/5
7. 1/5
8. 1/5
9. 1/10

p. 36
1. summer
2. 7
3. 27
4. 1/4
5. a. 7/27
   b. 2/9
   c. 10/27
   d. 4/27

p. 37
1. Tuesday
2. Wednesday and Friday
3. 30
4. a. 1/5
   b. 1/15
   c. 1/10
   d. 3/10

p. 38
1. 1/4
2. 2/3
3. orange juice, cherry or cranberry juice, apple juice, water
4. a. 4 gal
   b. 2 gal
   c. 1 gal
   d. 2 gal
   e. 3 gal

p. 39
1. 1 1/2
2. 2 1/6
3. 1 1/6
4. 2 1/3
5. 3 2/3
6. 4 2/3
7. 7 2/3
8. Answers will vary.
9. Answers will vary.

p.40
1. 1/6
2. 3/4
3. 5/6
4. 1/6
5. 5/9
6. 7/8

p. 41
1. cross out 4/7; 4/6
2. cross out 2/12; 3/12
3. cross out 4/30; 4/32
4. cross out 6/16; 8/16
5. cross out 10/12; 9/12
6. cross out 8/21; 8/28 or 6/21
7. cross out 25/28; 20/28
8. cross out 3/10; 3/12
9. cross out 6/8; 5/8
10. cross out 24/33; 27/33
11. cross out 10/30; 12/30
12. cross out 2/15; 2/10

p. 42
1. 5/10; 0.5
2. 80/100; 0.80
3. 1 32/100
4. 3.2
5. 3 5/100
6. 70/100
7. 0.54
8. 0.04
9. 5.8
10. 2
11. 0.2

p. 43
1. No; The whole number was added to the numerator before dividing.
2. The fraction needs to be renamed as a decimal before the whole number is added.
3. 4.6
4. 7.82
5. 8.43
6. 5.88
7. 12.3
8. 9.17

p. 44
1. a. 8
   b. 10
   c. 8 to 10; 8:10, or 8/10
2. a. 9
   b. 1
   c. 9 to 1; 9:1; 9/1
3. b
4. a
5. c
6. a

p. 45
1. 1/2
2. 3/10
3. 1/4
4. 14/25
5. 23/100
6. 19%
7. 32%
8. 71%
9. 2%
10. 20%
11. 25%

p. 46
1. 1/100; 0.01
2. 1/20; 0.05
3. 1/10; 0.10
4. 1/4; 0.25
5. 56/100; 0.56
6. 1 35/100; 1.35
7. $2.50
8. $6.67